MW01295294

Leif Erikson

A Captivating Guide to the Viking Explorer Who Beat Columbus to America and Established a Norse Settlement at Vinland

Free Bonus from Captivating History (Available for a Limited time)

Hi History Lovers!

Now you have a chance to join our exclusive history list so you can get your first history ebook for free as well as discounts and a potential to get more history books for free! Simply visit the link below to join.

Captivatinghistory.com/ebook

Also, make sure to follow us on:

Twitter: @Captivhistory

Facebook: Captivating History:@captivatinghistory

Contents

Introduction

The formidable sight of giant, blond, sword-wielding Norsemen sailing up river in their dragon boats to attack, raid, and pillage was enough to strike fear into the heart of any European community or settlement. The Vikings were a fearsome nation, conquering all who lay before them as they expanded their influence and power throughout Europe and destroying those who tried to halt their expansion across the continent. These terrifying Viking warriors were feared throughout Europe for their brutal and ruthless attacks, but they were so much more than fearsome warriors.

The Vikings were also exceptional boat builders, seafarers, adventures, and explorers. They not only raided and attacked settlements across Europe and Britain, but they also sailed far and wide, discovering and colonizing new lands. Their impact on history is far-reaching and the mark that they made on the world, especially during the Viking Age, can still be seen today.

The Viking Age was a time of expansion, conquest, and exploration and as a result, Norse culture and ancestry are found throughout Europe and the world. During this time the Vikings were sailing far and wide, discovering new lands and settling new territories. They were not afraid to take on their enemies or harsh climates to expand their territory and influence. This was the culture that famous Norse explorer Leif Erikson was born into. This was the world that shaped him and made him into an adventurous man who was not afraid to sail into the unknown to discover new lands. It was this background and heritage that enabled Leif Erikson to cross the Atlantic on an open longboat and become the first known European to have set foot on the North American continent, almost half a millennium before Christopher Columbus.

Chapter 1 – The Making of the Man, Leif Erikson's Formative Years

Over the centuries, much has been written about Leif Erikson and his remarkable voyage across the Atlantic, but unfortunately, no historical records exist from his lifetime, and most accounts of the discovery of Vinland are based on two Icelandic sagas written centuries later. His story has been embellished, diminished. and altered depending on the motivations of the writer. As a result, the man, the myth, and the legend have become inseparable, making it difficult to establish where the facts end and the fiction begins. But even if one cannot entirely separate fact from fiction, there is no denying that Leif Erikson made a significant impact on the world. This incredible Norse explorer not only changed the face of his world by discovering and exploring new territories, he also changed the land he lived in by converting the Greenlanders to Christianity.

No great events in history happen in isolation, and to understand Leif Erikson and his discovery of the North American continent, one needs to understand the world he grew up in. To do this effectively, it is necessary to take a closer look at what was happening in and around Europe during Leif's formative years. Leif Erikson was, without a doubt, a product of his time and upbringing. The legacy of his ancestry would have had a profound effect on the development of his character. His family, environment, culture, and the events taking place around him in his formative years would all have shaped the man he was to become.

We've all heard stories of mighty Viking warriors sailing their longboats far and wide to conquer new territories. Children around the world grow up being told about these formidable Norsemen who were famous for plundering, raiding, and looting settlements across Europe and Britain. We know the legends of Odin, the god of war, and his even more famous son, Thor, the god of thunder, and his mighty war hammer, Mjolnir. But the impact of the Vikings on world history goes far deeper than legends, war, and destruction.

The Viking Age (approximately 793-1066 AD) was not an age of war and destruction; it was an age of exploration, colonization, and settlement. It was a time when young, courageous Norsemen from across Scandinavia ventured forth from their homelands to make their mark on the world. During this time, Scandinavians could be found throughout most of the known world, from the Middle East to the distant shores of North America. Still today, the Norse legacy is seen throughout Europe.

The Swedes predominantly went east to Russia and the surrounding areas, while the Danes explored the North Sea coastline, England, and France. But it is the Norwegian voyages that are central to Leif Erikson's story. Some Norwegians went south like the Danes, but mostly they crossed the North Sea to conquer and rule Britain or sailed west across the Norwegian Sea and northern stretches of the Atlantic Ocean. The Norsemen plundered and pillaged their way through Shetland, Orkney, the Hebrides, Scotland, Ireland. and the

Faroe Islands as they ventured further and further from Norway until, toward the end of the ninth century, they reached the cold and distant shores of Iceland.

The first Vikings arrived on the shores of Iceland in approximately 860 as part of an exploratory party, and they did not settle there at the time. The island was named Iceland by Floki Vilgeroarson who was disappointed by the harshness of the environment. But despite this unfavorable first impression, the settlement of Iceland went ahead and the first Vikings settlers arrived sometime around 870. More than half the initial wave of settlers appear to have come from around Bergen in Norway.

There is much speculation as to why these intrepid Norsemen left their homelands to seek greener pastures elsewhere. The reasons appear to be as many and varied as the men who made the voyages. Dreams of glory in battle, wealth, trade, and political ambition would all have been compelling reasons for a Viking to cross the often treacherous and icy North Sea to conquer existing settlements and colonize new territories. These new areas may have offered more resources, land, and freedom to adventurous Vikings. In the new settlements and outlying territories, there would also have been fewer laws and social constraints. But it was not only pull factors that drove the Viking Age of expansion, as there were also push factors. Overpopulation and tyrannical rule were common in Scandinavia at the time and would have driven many Scandinavians to seek new and better lives elsewhere.

While the Vikings had a reputation as fearsome warriors, most of the men and women living in Scandinavia during this time were farmers, craftsmen, or slaves, and their lives consisted of hard physical labor in a demanding and often brutal environment. They were just as vulnerable to raids as the settlements they attacked and as susceptible to malnutrition, famine, and disease as anyone else at the time. These were all factors that made the allure of a fresh start in a new land worth the risk to many Norse families and as a result, voyages of discovery and exploration were often followed by voyages of immigration and settlement. Of course, there was also the appeal of the honor and prestige that went with discovering and settling new territories. And then there were the men who were banished or outlawed from their homeland for criminal behavior and who were forced to find new places to settle.

One such immigrant and outlaw was Thorvald Asvaldsson, Leif Erikson's grandfather. He was banished from Norway and forced to find a new place to live after being found guilty of manslaughter. Fortunately for Thorvald Asvaldsson and his family, including his young son, Erik Thorvaldsson, there were plenty of other places to go. The family left Norway and settled in Hornstrandir in the newly colonized northwestern part of Iceland.

Erik Thorvaldsson, commonly known as Erik the Red, was born in Norway and was approximately ten years old when the family moved to Iceland. The immigration of this family to Iceland is central to the story of Leif Erikson because he is the son of Erik Thorvaldsson. Leif was born in Iceland in 970 and his formative years – like those of any Norse child in this unforgiving land – were hard. In a time when 30 to 40 percent of children didn't reach adulthood due to disease, attacks, and malnutrition, Leif's chances of survival would have been little better than any other Viking child. But fortunately for history, Erikson did survive and thrive in the harsh environment that he was born into.

To better understand Leif Erikson and the traits that he developed during childhood and later used to his advantage when he left his home in search of a mysterious land to the west of Greenland, one has to take a closer look at his family dynamic and the role of his father. Not only in the family but also in the settlement of Greenland. Erik's discovery and settlement of Greenland and Leif's later discovery of Vinland on the North American continent are intrinsically linked. Leif may well have been the man who discovered Vinland but one cannot separate his story from that of Erik the Red and the settlement of Greenland. As his father, Erik would have had an immense influence, both good and bad, over Leif and the man he was to become. Erik may have taught Leif to be a farmer, warrior, and seafarer but Leif also appears to have inherited his father's determination, courage, and adventurous spirit.

Growing up in the Viking Age as Erik the Red's son would have had a profound impact on Leif's character. By all accounts, Erik the Red was a formidable figure. He was a large man with his flaming red hair and beard, and he most likely had a temper to match. He would have been a fearsome sight at the head of a Viking raiding party, striking fear into the hearts of any coastal settlement. But Erik the Red's contribution to world history is not as a Viking warrior but rather as an explorer and the man famous for the discovery and settlement of Greenland.

The settlement of Greenland is a pivotal point in Leif Erikson's life story. Once the Vikings had settled in Greenland, it was only a matter of time before they made their way to the North American continent. The Davis Strait between Greenland and North American is only 250 nautical miles at its narrowest point and crossing this distance would not have posed much of a challenge for seafarers who were accustomed to the 1,500 nautical mile Norwegian Sea crossing from Norway to Greenland. The Vikings were accustomed to dealing with frozen seas and ice flows, and the dangers of the crossing would not have put them off. When one looks at the Viking Age as an age of exploration and colonization, it is almost inevitable that a Greenlander would eventually form a settlement in North American. It is probably also safe to assume that if it had not been Leif Erikson, it would have been another Greenlander.

Chapter 2 – The Settlement of Greenland

According to the medieval Icelandic sagas, Erik the Red was born in Rogaland in Norway and grew up in Hornstrandir on the newly colonized Iceland. He married Thjodhild, and they moved from Hornstrandir to Haukadel where Erik built a farm called Eiriksstadir (he might have received the land as part of his wife's dowry) and started a family. Erik and Thjodhild had four surviving children, three sons, Thorvald, Leif, and Thorstein and a daughter, Freydis. But Erik had a fiery temperament, and he did not always get on well with his neighbors, arguing with them often. During one such argument, his temper got the better of him and he killed a man named Eyiolf the Foul.

The Vikings may have been ruthless warriors who raided and pillaged settlements across Europe, but within their own society the rules of law were strictly applied.[1] When Erik the Red killed Eyiolf the Foul in 982, the wheels of Viking justice were set in motion and this led to Erik standing trial, being found guilty of murder, and outlawed from Iceland and Norway by the Thorsnes Thing for three years. Banishment was a serious punishment for any Viking, and Erik was forced to leave Iceland and his family or face being killed by another Viking. But this is the perfect example of turning adversity into opportunity.

During his banishment, Erik the Red sailed west across the Atlantic until he happened upon Greenland, and then he spent the rest of his banishment exploring this new territory. What Erik initially found was a land that was mostly covered with glaciers and ice fields and was considerably colder than Iceland. But when he sailed up the western coast, he reached a part of the island that was free of ice and similar to Iceland. According to the Saga of Erik the Red, he spent the first winter on Eiriksey, the second winter in Eiriksholmar, and he explored as far north as Snaefell and Hrafnsfjord. The island may have been cold and harsh, with no great forests, but there was plenty of sea life off the coast, pastures for cattle, and adequate wildlife on the island to sustain a settlement. Off the coast of Greenland, the Vikings could hunt walruses, seals, and whales and on land they could hunt foxes, bears, and caribou. This enabled them to make a living as traders and buy much-needed resources, such as timber for homesteads and shipbuilding, from Norway.

While Erik sailed off into the unknown and explored Greenland, Leif remained on Iceland with the rest of the family. He would have helped around the homestead, worked with the livestock, and planted and harvested crops while his mother took care of the family. The young Leif would also have learned how to hunt and fish and develop other skills that he needed to survive in a harsh and inhospitable environment. Fortunately, because of their position in society,[2] the family would have had thralls to help work the farm while Erik was outlawed. One man who had a great influence on Leif's young life, especially during the time his father was outlawed, was a man named Thyrker. Thyrker was a German who Erik the Red had captured and brought back to Iceland, but he does not appear to have been a slave. By all accounts, Thyrker was a good teacher, and he taught Leif everything he needed to know about weapons and battle tactics, but he also taught him how to read and trade. Even after Erik the Red had returned to Iceland, Thyrker continued to play an important role in Leif's life, and he accompanied Leif on his voyage to North America.

Erik was only partially outlawed, and this meant that he was allowed to keep his land and possessions and return to his family after three years. Erik did return to his family, but by then he had no intention of staying on Iceland, as he had seen the potential to start a new settlement on Greenland. The land may have been colder, less hospitable, and with larger glaciers than Iceland, but Erik managed to persuade a number of Icelandic Vikings that there was enough land to farm, there were pastures suitable for raising livestock, and the fjords were teeming with cod and other fish.

In the summer of 985, Erik the Red and his followers set sail for Greenland. This was no small undertaking. There were twenty-five ships, approximately 300 settlers, horses, cattle, sheep, goats, pigs, and various equipment. Making the journey from Iceland to Greenland was a treacherous and risky venture, and of the twenty-five ships that set sail from Iceland, only fourteen made it safely to Greenland. The rest either turned back or were lost at sea. During this voyage, and subsequent voyages that he made with his father between Greenland and Iceland, the young Leif would have learned valuable lessons about ocean crossings and deep-sea navigation. It was this knowledge that would have enabled him to make the crossing from Greenland to North America.

Making it from Iceland to Greenland was merely the beginning for the settlers, and once they landed on Greenland things did not get any easier. Every day was a constant struggle for survival. Homesteads had to be built, crops planted, the livestock needed grazing and shelter, and the men had to hunt and fish to feed their families. But Erik was a competent man, and under his leadership, the settlement flourished and grew as more immigrants braved the treacherous voyage from Iceland. Establishing a new settlement in an unknown and hostile land was a daunting venture and having the courage to attempt it and the determination to succeed gives one a clear indication of the type of people the Greenlanders were.

Erik the Red may be credited with discovering Greenland and starting the first settlement on the island, but he did not single-handedly settle Greenland. This was a joint effort that involved all the settlers and relied on everyone pulling their weight. By the time the first settlers arrived in Greenland, Leif was old enough to do his share of the work. and he would have toiled alongside the men to build homesteads for the families and shelter for the livestock. He would have experienced firsthand what it took to settle a harsh and inhospitable land.

This early introduction to settling a previously uninhabited territory and the knowledge he gained working alongside his father would undoubtedly have stood Leif in good stead when he later landed on the shores of North America. He would have known how to choose a suitable site for settlement and what tasks needed to be completed in order to establish a base camp in a new territory.

His early life on Greenland would have given him the confidence, skills, and knowledge he required to strike out on his own. By all accounts, Erik the Red sailed frequently between Iceland and Greenland to trade, and Leif made many trips with his father. Erik taught his son about ocean crossings and navigation and apparently, Leif had a natural aptitude for sailing and developed a reputation as a seafarer. One legend tells how Leif, when he was about sixteen years old, spotted a polar bear on an ice flow and decided to hunt the bear. He used his knowledge of the sea and currents to take his boat upstream from the bear and let the current carry him into the ice flow, and then he used the same method to get back to land. Apparently, the people watching from the shore were impressed by this tactic. There is no way of verifying the truth behind this story, but if it is true, then it is a good demonstration of Leif's understanding of the sea and its currents and his natural talent as a sailor. It also demonstrates his bravery and ability as a hunter. Clearly, Leif was not afraid to take risks to achieve his goals.

[1]*Viking law was well established and was the root of their government. The Vikings did not have a central government but the rule of law, fairness, and justice was important to them. They lived in an ordered and structured society made up of small clans and each clan had a chief. Decisions regarding village life were made at a "Thing." The Thing can be compared to a legislative assembly. The chief of the settlement ran the Thing but he and his council could only guide the Thing and all free men had the right to take part in the decision-making process. They voted on many things including, who owned a piece of land or what punishment a person would receive if they were found guilty of breaking the law.*

The Vikings had no law books and the laws were not written down, but one person in every village was assigned the duty of law-speaker, a very important position in a settlement. Some laws applied to all Vikings but others were specific to the settlement, and the law-speaker had to know them all. He had to memorize the laws of the settlement and the broader Viking laws. If there was any confusion about a law during a Thing, the law-speaker would explain the law. After new laws were passed at a Thing, the law-speaker had to go to the law rock and recite all the laws so that the women, children, and slaves, who were not present at the Thing, would also know the laws.

The Thing was not only a time to make new laws but also to enforce the law and try men accused of committing crimes. During the Thing the accused could defend themselves and call witnesses. The rule of law was important to maintain order in the harsh and often violent Viking settlements and punishment could be severe. The ultimate penalty for violent crimes was usually to outlaw or banish the guilty. Outlaws were banished and had to flee their village and hide in the wilderness because anyone was allowed to hunt down and kill an outlaw. Outlaws also lost all their worldly possessions and property. Depending on the crime, the banishment could be permanent or for a set number of years. If a Viking was only outlawed for a certain time, they were usually allowed to keep their possessions and return to their homes when they had served their time. Lesser crimes could be settled by a fight, a holmgang or duel. This did not always end well for the victim, and justice was not necessarily on the side of the right, but since the Vikings believed the gods favored the righteous, the outcome was seen to be fair justice. Land could also be confiscated from the guilty and given to the victim.

Once a year, the clan gathered for an Allthing where all free men in a settlement were able to vote on important decisions, like taxes, peace treaties, the election of a new chief, or the adopting of new laws.

[2]*Norse society basically had three tiers. Jarls were the nobility, and Kings were drawn from this stratum of society. In Norse society, the position of king was not always guaranteed through succession. The jarls could unseat a weak or unpopular monarch if they united behind a rival claimant. In peacetime, the jarls oversaw the running of their lands, and during war and raids they commanded the*

longboat crews. The jarls venerated the god Odin for his wisdom and knowledge. The karls, or Norse middle class, were farmers, fishermen, and craftsmen. On raids or during wars, they crewed the longboats and were the rank and file of the Viking army. The kuskarls (house karls) served a jarl or king on his personal staff or as bodyguards. Thor was a principle deity of this class and was venerated for his honor and bravery as a warrior. The third strata of Norse society were the thralls, who were little more than slaves, convicted criminals, and captives from raids. Thralls had no rights and the killing of a thrall was considered destruction of property rather than murder. Viking raiders bought and sold slaves and captives from parts of Europe, Britain, and the Middle East so the thralls were not one ethnic or cultural group and could come from anywhere.

Chapter 3 – Life in Greenland

There were two settlements on Greenland about 400 miles from each other. The Eastern Settlement (**Eystribyggð** at present-day **Qaqortoq**) and the Western Settlement (Vestribyggð which is close to present-day **Nuuk**). These were the only two areas that were suitable for farming and homesteads, and the settlements were widely dispersed so that there was enough land for each family to farm and enough grazing for cattle. The Book of the Settlements tells how Erik the Red built his family estate, **Brattahlíð**, in the Eastern Settlement at the head of the Tunulliarfik Fjord, approximately 96 kilometers from the coast. At the time, the Fjord was known as Eiriksfjord, and the area had some of the best farmland on the island because it was protected from the Arctic sea and cold, foggy coastal weather.

The settlers on Greenland built typical single roomed, low, rectangular-shaped Viking dwellings. The only difference between the homesteads in Scandinavia and those on Greenland was the building material. There were no great forests on Greenland and timber was scarce, so the walls of the houses were made of stone rather than wood. The remains of typical houses and farm buildings found on Greenland had stone walls, approximately 1.5 meters thick and covered with a turf outer bank to provide much-needed insulation. The houses also had flagstone floors rather than the more traditional earth covered ones with reeds.

The interior of a Norse house was plain and basic. A single room served as both living and sleeping area for the average Viking family. Reeds or flagstones covered the ground and damp rose up through the rough walls while the wind whistled through the small openings that served as windows. The smoke from the central cooking hearth swirled around the dark interior of the dwelling before escaping out of a hole in the reed roof. The lodgings may have provided shelter from the harsh weather, but there were few comforts. There would have been a table and a few stools where meals were eaten and rug-covered benches along the walls that served as beds. Eric the Red's dwelling may have had what were considered a few more luxuries than your average Viking, but it certainly wouldn't have been grand or lavish by today's standards. The house had more than one room and there might have been a tapestry or two hanging on the walls. The family probably slept on straw-stuffed mattresses rather than hard benches, but the house would still have been dark, damp, and smoky.

The image of the Vikings as bloodthirsty warriors crossing the seas on their terrifying longboats to loot and pillage without mercy is one that has endured throughout the ages, but your average Norseman spent most of his life farming, trading, or working as a craftsman. The men were blacksmiths, fur traders, hunters, farmers, fishermen, and shipbuilders. They worked the land and grew crops such as wheat, barley, and rye and they tended to their livestock. The Greenlanders took sheep, goats, pigs, cattle, geese, and chickens with them.

During the summer when the weather improved around Greenland, each settlement would send men to hunt in Disko Bay above the Arctic Circle. They returned not only with meat that could be dried and eaten during the long winter months when fresh food was scarce, but also other valuable commodities such as seal pelts and walrus tusks which could be traded and sold so that they could buy much-needed timber from Norway and other supplies that they could not produce for themselves.

The women took care of the family; they raised the children, made the clothes, and cooked the meals. The food was basic, and they had to make do with what they could grow themselves or gather from the surrounding area. The Viking diet consisted mainly of bread, porridge, cabbage, onions, leeks, and wild berries. While to some modern people this subsistence lifestyle may sound like some kind of romantic idyll, it was anything but. It was a hard, perilous existence where daily life was often a grueling struggle for survival. The men may have been hunters and warriors, but the women were just as tough and had to look after their families when the men were away hunting or raiding.

But the life of a Viking settler was not all work and drudgery, there was also time for sport and other entertainment. The Vikings enjoyed wrestling, ice-skating, skiing, archery, and falconry. They demonstrated their strength with stone-lifting competitions and enjoyed a game called knattleikr that involved a ball, full body contact and, at times, a bat. These games would surely have provided a welcome respite from working the lands and caring for the livestock. But the games, which were often violent and could result in serious injury or even death, were not just pure entertainment; they also allowed the men to practice vital battle skills and improve their fighting techniques. The games also enabled young boys to learn the skills that they would need to be great Viking warriors and to demonstrate their strength, agility and battle tactics. Leif Erikson would no doubt have taken part in many different Viking sports and games and been as eager as any other young Viking to show off his strength and battle skills to impress his father.

Norsemen were accustomed to spending much of their time outdoors, but during the long, harsh winters they would have been forced to spend time indoors, and board games were a popular pass-time. They amused themselves playing dice, chess, and board games like Hnefatafl and Kvatrutafl. Kvatrutafl was a game similar to backgammon, and Hnefatafl was a type of war game that is thought to have helped teach players battle strategies. The Norsemen also enjoyed listening to stories, telling riddles, and playing musical instruments such as harps, horns, and pipes.

Storytelling and the reciting of sagas, especially those about the exploits of Viking heroes, were deeply ingrained in Norse society. During Viking feasts, Scandinavian bards called skalds[3] recited epic poems or sagas that praised the brave deeds of Viking warriors and their prowess in battle. These sagas were important to the Vikings because before they converted to Christianity they had no written records of significant events, and their history was passed on through the sagas told by the skalds. This was their only way of preserving their history, and these sagas were often incredibly long and detailed. Some of the sagas, like the Saga of Eric the Red and the Saga of the Greenlanders, were eventually written down long after the events that they describe, but many were lost to history.

[3]*One of the best-known Icelandic skald's is Snorri Sturluson. Snorri was born in Iceland in 1179 and became a wealthy and renowned poet and law-speaker. He wrote the Prose Edda, which, together with the Poetic Edda (written by Saemunder Sigfusson), has given modern historians much insight into Norse mythology. As a poet,*

Snorri would have heard many legends and tales that at the time only existed in oral form and may have decided to preserve these tales in the Prose Edda. He also composed the Heimskringla, a history of the kings of Norway from the ninth century to 1177. Snorri Sturluson is a good example of the importance of the skald in Scandinavian society. Without skalds, much of the history and culture of the Vikings would have been lost. It was skalds like Snorri Sturluson that kept the memory of Leif Erikson and his discovery of Vinland alive for future generations.

Chapter 4 – Leif's First Voyage

Leif was born to be a seafarer and adventurer. It was in his blood and part of his heritage. Leif's early introduction to sailing, combined with regular crossings from Greenland to Iceland, gave him the seafaring and navigational skills that he needed when he set out with a small crew in search of the mysterious lands that he had heard lay to the west of Greenland. But it was not only seafaring skills that he acquired from his father, it was also confidence and determination. Leif probably believed that if his father could discover and settle a new land, there was no reason that he could not do the same. He may have wanted to emulate his father's success or prove that he was as good of an explorer and leader. Or perhaps it was a desire for adventure and wealth that drove Leif to sail west into the unknown. Or maybe it was as a combination of all these factors. Too much time has passed for us to ever know what motivated Leif, but it is impossible to ignore the fact that he was the son of a great Norse explorer and leader, and he obviously inherited many of his father's traits.

Erik the Red was obviously a natural leader. He persuaded 300 people to leave their homes and sail across the treacherous ocean to settle an unknown, but by all accounts, inhospitable, land. Those who joined Erik's colonizing party obviously had many and varied reasons for leaving Iceland, but the fact that he was able to persuade so many to go with him is testament to his competence and leadership. Clearly, these people trusted him and were willing to follow him to a land that they had never seen because he persuaded them that it could offer them a better life. When the settler party reached Greenland, Erik was elected paramount chieftain of the colonists in the Eastern settlement where he established his homestead. His position ultimately gave him great wealth and power, but settling this new territory was by no means easy. Building homesteads and shelters before the onset of the first harsh winter, hunting game to feed the settlers. and finding enough grazing for the livestock would have been all-consuming.

The daily struggle against a harsh environment bred hard men, and Leif was no exception. He grew up surrounded by hard, often violent role models. His father and grandfather were no strangers to violence, and both had murdered fellow Norsemen and been outlawed for their crimes. Both Erik the Red and Leif Erikson left their homelands when they were young and traveled with their families to settle in new territories. These early experiences would undoubtedly have had an effect on their characters and taught them how to survive and thrive in the harsh conditions of a new settlement. From an early age, Leif would have learned the skills he needed to be an effective colonist. He would have learned how to be a successful hunter, fisherman, seafarer, and warrior but more importantly than that, having grown up surrounded by adventurous men and women who were not afraid to strike out on their own, explore new territories, and establish settlements in hostile environments, Leif Erikson would probably have been encouraged to make his own mark on the world. He certainly would not have wanted to spend his entire life in his father's large and imposing shadow.

At the age of 24, Leif was ready to captain his first voyage, and he went eastward to the homeland of his family. Around 997 AD, he set sail for Norway with a crew of fourteen men (including Thryker), bearing gifts for King Olaf Tryggvason. At the time, it was not unusual for young men to serve as retainers in the household of a king or chief, and this was an opportunity for Leif to forge alliances that would gain him status and political influence that could help him in the future.

But Leif's first voyage was not all smooth sailing. It started off well enough. The weather was fair and the wind was good when Leif set sail from Greenland. Unfortunately, after their first day at sea, the wind died down and it took them five days to reach Iceland, a voyage that usually only took two days. Initially, the crew wanted to go ashore at Iceland, but Leif decided to bypass the island and continue sailing toward Norway. They sailed on, and when they finally sighted land again, they realized that the land they could see was the Hebrides. They had sailed further south than they had intended.

Leif and his crew had to go ashore on the Hebrides to restock their boat with food and supplies. While they were still on the island, a storm rolled in and the boat could not leave the Hebrides for more than a month. During this time, Leif stayed in the house of the chief of the island and there, his striking appearance apparently caught the attention of the local chief's daughter, Thorgunna. Before Leif left the Hebrides, he conceived a son, Thorgils, with Thorgunna. Although Leif acknowledged his son's paternity, he and Thorgunna never married, and Thorgils spent most of his childhood in the Hebrides with his mother. Later, he was sent to Greenland to live with his father, but by all accounts, he was never very popular with the Greenlanders.

After the storm cleared, Leif left the pregnant Thorgunna in the Hebrides and continued his voyage to Norway. On this leg of his voyage, the wind was favorable, and it only took a few days to make the crossing from the Hebrides to Norway. When he arrived in Norway, he was welcomed to the court of King Olaf who knew his father. King Olaf was another powerful and adventurous man who had a significant influence on the life of Leif Erikson.

The king was clearly impressed with the young Leif and invited him to spend the winter with him in Norway. Leif accepted this generous offer and by all accounts enjoyed his time in the Norwegian monarch's court, enjoying the luxuries that it had to offer. The time that Leif spent with King Olaf was to have a lasting impression on the young man. One of the greatest impacts that King Olaf had on Leif's life was his conversion to Christianity. Until that time, Leif, like the rest of his family and the other settlers on Greenland, had worship pagan Norse gods.[4]

Little is known about the birth and early life of King Olaf Tryggvason, but he became a mighty Viking warrior who acquired great wealth and fame by raiding and plundering throughout Europe, and he was the first Christian king of Norway. King Olaf was not born a Christian. Being a Norseman, he would have been raised to worship the same pagan gods that the Greenlanders and many other Norwegians worshipped at the time, but he later converted to Christianity. King Olaf probably converted to Christianity during his campaign in Britain around 994, and by the time he bought the Christian faith to Norway, most of Western Europe was already Christian.

In 996, King Olaf returned to Norway and after a Thing, he was proclaimed king. As king, he set about converting all of Norway to Christianity, and he did so in a brutal manner. Pagan temples were destroyed, and churches were built in their place. Those who refused to convert were killed, tortured, maimed, or banished. After bringing Christianity to Norway, King Olaf was determined to spread his religion to the outlying areas of Iceland and Greenland. In Iceland, Christianity was adopted at an Althing in the year 1000 (after the Greenlanders had already left the island). Conversion in Greenland was never mandated by an Althing and was a more gradual process.

In order to spread Christianity to the distant shores of Greenland, King Olaf needed a convert with influence in the settlements and in Leif, he found just such a person: a young man who he was able to convert to the faith and task with spreading Christianity to Greenland. Before Leif and his crew returned to Greenland, they all converted and were baptized in their new faith. According to Erik's Saga, King Olaf set Leif the task of converting his fellow Greenlanders to Christianity and even sent a priest back to Greenland with him.

Leif took his mission from King Olaf seriously, and on his return set about teaching his fellow Greenlanders about his new faith and converting them to Christianity. Like his father, Leif obviously had the ear of the people and knew how to persuade them to his way of thinking. Using his influence and reputation as a man of fair judgment and honesty, Leif was able to convert many Greenlanders, including his mother, Thjodhild, to his new faith. Leif's father, however, was a different matter, and he clung to his old Norse ways and continued to worship his pagan gods. No matter how hard he tried, Leif was never able to convert his father to Christianity, and Erik the Red died a pagan. But Thjodhild became an avid supporter of the Christian faith and records indicate that she commissioned

Greenland's first church. In 1932, a group of Danish archaeologists excavating Brattahlid (Erik the Red's homestead) found the remains of what they assume is Thjodhild's church. As a preeminent Norse seafarer and explorer, Leif Erikson redefined the Viking world, and his conversion to Christianity changed the beliefs of the society he lived in. These are remarkable achievements and show both the character of Leif and the standing that he had in the community.

[4]*Religion and beliefs were very important to people throughout Europe and, like most pagans, the Norsemen were superstitious and worshipped many different gods. In the Viking view of the world, there was Asgard, the home of the gods. Asgard was a huge fortified castle that floated in the air. At the edge of the world lived the giants, and the Vikings equated them with chaos and destruction. They were related to the Nordic gods and battled with them continuously. In the Nordic legends, Thor often went out to hunt giants. The Norsemen believed that at the end of the world, the gods, giants, and humans would meet in one final deadly battle. The pagan Norsemen worshipped gods like Odin, Freya, Thor, and many others. In Norse mythology, Odin is the ruler of the deities, a war god and a seeker of wisdom, and Freya is his wife. Thor, who, thanks to popular media, is probably the best known of the Norse gods, is Odin's son and the god of thunder. Thor is an embodiment of many qualities that are traditionally associated with the Vikings. He is a mighty warrior, a fearless traveler, and his courage, strength, and loyalty are legendary. Leif would have spent many long winter days in his parents' dark, smoky homestead listening to tales of the exploits of Thor, and he would probably have aspired to emulate this Norse god. Another deity who would have been of great significance to Erik the Red, Leif, and other seafaring Norsemen was Aegir, who was loved and feared in equal measure as the commander of the sea. Aegir and his wife, Ran, were said to dwell in*

a magnificent hall beneath the ocean. While Aegir is portrayed as a gracious host, Ran is depicted as drowning unfortunate sailors and dragging them down to dwell in the magnificent hall beneath the ocean. Whenever a Norse ship sank, it was said that the sailors were dining in Aegir's Hall, and Norwegian seafarers had a deep respect for the ruler of the seas.

Chapter 5 – Mysterious Lands to the West of Greenland

King Olaf himself had set Leif the task of converting the Greenlanders to Christianity, and by all accounts he took his mission seriously, but even so, Leif did not remain on the island for long after he returned from Norway. Having been raised by an adventurous father, Leif had clearly developed an adventurous spirit of his own, and before long, he felt the urge to travel and explore. Maybe his voyage to Norway had awakened a wanderlust in the young Norseman, or perhaps Leif found the drudgery of life on Greenland boring, or maybe his ambitions were stifled by his father. We will never know for certain what drove Leif to leave Greenland and go in search of a land that he wasn't even sure existed, but what we do know is that shortly after returning from Norway, Leif embarked on the greatest adventure of his life. A voyage that would make him famous, not only in his in own time, but would also ensure his place in world history.

Like most great historical events, Leif's discovery of Vinland did not take place within a vacuum. It was not an isolated incident or random turn of events that led this renowned seafarer and navigator to the shores of the North American continent. The story of the discovery of Vinland began many years before Leif made his epic voyage.

For many years, the honor of being the first European to set foot in North America was given to Christopher Columbus, the Italian explorer, navigator, and colonizer, who landed on the continent in 1492. While the contribution that this great European explorer made to history cannot be underestimated, it is now widely acknowledged that he was not the first European to discover the continent. That honor has now been placed squarely on the broad shoulders of Leif Erikson and today, few historians dispute the fact that Leif beat Columbus to North America by almost 500 years.

Historians may well agree that Erikson discovered North America, but unfortunately, they still cannot reach a consensus on whether this discovery was on purpose or by accident. Most historical knowledge of Erikson's discovery of North America comes from sagas and other ancient sources, written decades after his voyage. There are no contemporary accounts of Erikson's voyage or the settlement he established at Vinland.

There are, however, two main sagas that relate to the discovery of Vinland. These are the Groenlendinga Saga (Saga of the Greenlanders) and Eirik's Saga,[5] and both were written in Iceland approximately 250 years after the Vikings had been to Vinland. While both sagas were written at much the same time and both mention Leif's voyage across the Atlantic, that is where the similarity ends, and the two accounts differ considerably.

According to Eirik's Saga, Leif's sighting of the North American continent was entirely accidental. In Eirik's Saga, Leif Erikson sails from Greenland to Norway and stays in King Olaf Tryggvason's household for the winter, but on his return journey he is blown off course, and when he eventually sights land, it is not Greenland but some other unknown landmass. The land he stumbles upon is nothing like Greenland, with its glaciers and harsh climate. Leif later describes the land as green and fertile, with dense forests and an abundance of wild grapes. Leif names the land Vinland (most likely because of the grapes) and he and his crew establish a small village where they can spend the winter in relative comfort. After the winter, Leif and his men sail back to Greenland and never return to Vinland. After that Leif, almost entirely disappears from the saga and the Icelander Thorfinn Karlsefni becomes the hero of the story. It is Thorfinn who sets out on an expedition to found a settlement in North America.

The Groenlendinga Saga tells a different tale. In this version of events, it is not Leif who is blown off course and stumbles across the North American continent but rather a man named Bjarni's Herjolfsson. According to the Groenlendinga Saga (Saga of the Greenlanders), Bjarni Herjolfsson was an Icelandic trader and the son of Herjolfr Bardarson. Bjarni traded between Norway and Iceland and was by all accounts a very able seafarer and navigator. When he returned to Eyrarbakki, Iceland, from one of his trading voyages, he discovered that his parents had left the island with Erik the Red to settle on Greenland. Bjarni decided to follow his parents to Greenland but never having been to the island before; he had to rely on his considerable navigational skills and a description of Greenland to find the place.

He and his crew were blown off course by stormy weather, and the first land they sighted was nothing like the description they had of Greenland. What they saw was a land that was hilly, fertile, and covered in great forests. As Bjarni and his crew sailed further north along the coast, they did see some snow-covered mountains but no great glaciers and fields of ice. Bjarni quickly realized that this could not possibly be Greenland, and he changed course and sailed on without setting foot on land. When he eventually made it to Greenland, he mentioned the land he had seen. At the time, none of the newly settled Greenlanders went in search of this land, and it was only years later that Leif made his voyage. If one assumes that Bjarni was blown in a southwesterly direction from Iceland, then it was probably Newfoundland, Baffin Island, and the Labrador Coast that he had seen.

Because no contemporary accounts exist of Leif Erikson's discovery and settlement on the North American continent, it is impossible to know which account is more accurate. But if one considers that Leif Erikson didn't act in isolation, then it appears to be more likely that he set out to find the mysterious land that Bjarni had seen to the west of Greenland rather than stumbled upon it. Leif Erikson was a product of his time. He grew up during the Viking age. He was an accomplished seafarer and warrior, and his father was a respected Viking explorer who discovered and settled Greenland. If you consider all these facts, it is highly likely that his discovery was a deliberate act rather than a navigational miscalculation or accident.

Another part of the story that leads one to believe that it was a planned expedition is the notion that Erik the Red was originally going to accompany his son on the voyage. Unfortunately, Erik was ruled by his superstitions and legend has it that he fell off his horse shortly before they were due to set sail. Although not seriously injured, he saw it as a bad omen and decided not to join the expedition. This may have been unfortunate for Erik, but maybe not so bad for Leif. Who knows how history would have recorded the story if Erik the Red had accompanied his son across the Atlantic? Maybe Erik would have been credited with being the leader of the expedition and the discovery of the North American continent, and

Leif would have been relegated to a mere footnote in history. Perhaps it would have been Erik's achievements that are celebrated throughout America and his statues erected in numerous American cities. Erik the Red was undoubtedly an ambitious man and a natural leader, and it is hard to see him standing back and allowing his son to take all the glory.

When Bjarni reached Greenland, he settled there with his parents, and when his father died, he inherited his estate and spent the rest of his life on the island. Bjarni arrived in Greenland approximately fifteen years before Leif Erikson made his historic voyage to North America. Growing up on Greenland, Leif would no doubt have heard the stories of the mysterious land Bjarni had seen to the west. Some scholars suggest that Leif may even have heard the stories directly from Bjarni and, considering that the settlement on Greenland was not particularly large, this is highly possible. But even if Leif had not heard the stories directly from the Icelandic trader, he would have known about the land, and he would have heard descriptions of the thick forests along the coast. This would most definitely have piqued Leif's interest because, while Greenland had grazing for livestock and land to farm along the fjords, it did not have large forests, and timber was in short supply on the island. Timber was a vitally important commodity to the Vikings, especially for shipbuilding. A reliable source of timber would have been highly sought after by the settlers.

Timber was central to the Norse way of life, and it enabled them to become the powerhouse they were during the Viking Age. It was necessary for building houses, but just as importantly, it was essential for building Viking longboats. Norway's vast pine forests combined with Norse shipbuilding skills allowed them to not only raid other settlements but also explore and settle new lands. A source of good timber was vital to their age of exploration and expansion but, unfortunately, there were no great forests on Iceland or Greenland. Before Leif traveled to North America, the settlers had to import timber from Norway to build longboats or, alternatively, they would have bought their longboats from Norwegian boat builders.

The Norse settlements on Greenland and Iceland were predominately farming settlements, but ships were still central to their survival in these far-flung regions. Not only did the ships allow the settlers to trade with other settlements, but having good, strong longboats meant they could fish and hunt whales and seals in Disko Bay above the Arctic Circle.

[5]*The Sagas*

Much of what is known about the history of Greenland and the discovery of Vinland come from two Icelandic sagas, namely the Sage of Erik the Red and the Saga of the Greenlanders. Both sagas are based on stories persevered for generations by oral traditions and were written in Iceland approximately 250 years after the events they describe. While both sagas make reference to Leif Erikson and his voyage to the North American continent and contain similar elements, they also differ greatly. Here is a brief summary of the two sagas.

The Saga of the Greenlanders (Groenlendinga Saga)

The Saga of the Greenlanders not only describes Leif's voyage to Vinland but also subsequent voyages to the new land. The Saga of the Greenlanders describes haw Bjarni Herjolfsson accidentally discovers a new land to the west of Greenland in about 986 and how fifteen years later, five expeditions are made to Vinland, although one voyage, led by Leif's brother, Thorstein, fails to reach its goal.

In this version of the discovery of Vinland, Bjarni does not go ashore. He only sees the landmass from his boat but when he realizes it cannot possibly be Greenland because there are no glaciers and ice-flows, he sails on without making landfall. When Bjarni finally reaches Greenland, he settles on his father's farm and there he remains. He never attempts to mount an expedition to return to the mysterious land that he had seen. He does, however, tell his story to other settlers on Greenland, and this ultimately inspires Leif Erikson to organize an expedition to explore this land some fifteen years later. In the Saga of the Greenlanders, Leif is the main explorer of Vinland, and he establishes a base camp at Leifsbudir. This serves as a boat repair station, storage area for timber and grapes before they are shipped to Greenland, and a base for subsequent expeditions.

Leif retraces Bjarni's route in reverse past Helluland (land of flat stone) and Markland (land of forests), and then he sails on across the open sea for another two days until he finds a headland with an island just offshore and a pool accessible to ships at high tide. Leif and his crew made landfall in the area and established a base. They name the place Vinland and the winter is described as mild rather than freezing, and it is in this area that Thyrker is reputed to have found an abundance of wild grapes. In the spring, Leif returns to Greenland with a boatload of timber and grapes. Leif never returns to Vinland. The second expedition to Vinland is led by Leif's older brother, Thorvald, with a crew of about 40 men. This group spends three winters at the base that Leif had established and named Leifsbudir. They explore the west coast of the new land in the first summer, and the east coast in the second summer.

During their exploration of the area, they make contact with the local inhabitants that they called Skraelings, and violence breaks out. After killing some Skraelings, the Norse explorers are attacked by a large force, and an arrow fatally wounds Thorvald. The following spring, the remaining Greenlanders decide to return home. Leif's younger brother, Thorstein, leads a third expedition to Vinland to recover Thorvald's body, but he is driven off course and spends the summer wandering aimlessly in the Atlantic before returning to Greenland, having failed in his mission. The following winter, Thorstein dies from illness, and his widow, Gudrid, marries Thorfinn Karlsefni, an Icelander. Thorfinn agrees to lead another expedition to Vinland. This is a larger expedition, and Gudrid accompanies her husband, taking livestock with them. Gudrid gives birth to a son, Snorri, in Vinland, but shortly after his birth, the group is attacked by the local inhabitants. They manage to retreat to a defensive position and are able to survive the attack. The following summer, they return to Greenland with a cargo of grapes, timber, and hides. Shortly after this, Leif's sister, Freydis, persuades the captain of an Icelandic ship to mount an expedition to Vinland. They set sail in the autumn and spend the winter at Leif's camp, but a disagreement leads to the killing of the Icelanders. The Greenlanders then return home with their cargo. This is the last Vinland expedition recorded in the Saga of the Greenlanders. According to the Saga of the Greenlanders, Leif's voyage to Vinland was planned and deliberate.

The Saga of Erik the Red (Eirik's Saga)

In this version of the story, Leif accidentally discovers the North American continent on his return to Greenland following a visit to King Olaf Tryggvason in Norway. Leif spends a winter in the household of King Olaf where he is converted to Christianity. King Olaf then commissions Leif to spread Christianity to Greenland and convert the Greenlanders to the faith. On his return voyage, he is blown off course during a storm and makes landfall on a mysterious land where he spends the winter. On his return to Greenland, he brings with him not only the Christian religion but also a cargo of grapes, wheat, and timber. He also rescues survivors from a wrecked ship, and this earns him the nickname Leif the Lucky, and his religious conversion of the Greenlanders is a resounding success.

The Saga of Erik the Red, like the Saga of the Greenlanders, states that this was the only voyage that Leif made to Vinland. In the spring after Leif returned, his younger brother, Thorstein, leads the next expedition to the new land but is driven off course by a storm and spends the entire summer wandering aimlessly in the Atlantic. He returns to Greenland without ever making it to Vinland. On his return, Thorstein marries Gudrid, but he dies of illness in the winter. The following winter, Gudrid marries a visiting Icelander named Thorfinn Karlsefni. He agrees to undertake the largest expedition to Vinland. His wife accompanies him on this voyage, and they also take livestock with them. They are accompanied by another pair of Icelanders, Bjarni Grimolfsson and Thorhall Gamlason, as well as Leif's older brother, Thorvald, his sister, Freydis, and her husband, Thorvard. They sail past Helluland and Markland and continue past some extraordinary long beaches before landing along the coast and sending out two scouts to explore the land. After three days, they return with grapes and wheat.

The expedition sails on until they come to an inlet with an island just offshore and there they make camp. This camp is called Straumfjord. The winter is apparently harsh, and food is scarce. When spring comes, Thorhall Gamlason wants to sail north to find Vinland, but Thorfinn Karlsefni wants to sail southward. Thorhall takes nine men and sails north, but his vessel is swept out to sea and never seen again. Thorfinn and the rest sail down the east coast with approximately 40 men and establish a camp on the shore of a lagoon. The settlement was known as Hop, and there they found an abundance of wild grapes and wheat. How long they stay there is unclear. They have contact with the local people (Skraelings). The first encounter is peaceful. Later, they return and trade with the Norse. One day, the local people are frightened by the Greenlanders' bull and they attack the explorers. They manage to survive the attack by retreating to a more defensive position. After that, the explorers abandon their southern camp and sail North again. Karlsefni and Thorvald Eriksson take a crew and sail in search of Thorhall. They once again have a hostile encounter with the local people, and Thorvald is shot with an arrow before dying from his wound. The explorers remain on the continent for one more winter, but the situation is tense, and there are disagreements among them. The following summer, they abandon their venture and start the return voyage to Greenland. This is the last Vinland expedition recorded in the Saga of Erik the Red.

A notable difference between the two sagas is that in the Saga of Erik the Red, Leif's role has been reduced to that of accidental discoverer of Vinland, and Thorfinn Karlsefni is the main explorer of Vinland. Bjarni Herjolfsson's voyage fifteen years earlier is not mentioned. In the Saga of the Greenlanders, there are five attempted expeditions to Vinland over the course of a number of years, but in Erik's saga, there is only one mega-expedition after Leif discovers Vinland. The name Leifsbudir does not appear in the Saga of Erik the Red and instead there are two camps, Straumfjord (Fjord of currents) and Hop (Tidal Lagoon). Straumfjord is the main base where the explorers spend the winter. Hop is a summer camp where timber is cut and grapes are collected and then shipped to Straumfjord before being taken to Greenland. The reasons for the differences in the two sagas is unclear. Both were based on oral histories and written long after the actual events, so it could be as simple as two different interpretations with different authors placing different emphasis on different events. Bearing in mind that Icelanders wrote both versions, they might have had different agendas, and the writer of the Saga of Erik the Red may have wanted to make the Icelanders' contribution to the discovery of Vinland the significant part of the story. Therefore, Thorfinn Karlsefni role is greatly embellished, and Leif Erikson is only mentioned briefly. The details of the two sagas may differ greatly but the fundamental premise that Leif Erikson was the first European to land on the North American continent is common to both.

Chapter 6 – Voyage to Vinland

Around the year 1000, Leif gathered a small group of around 35 men, including his mentor, Thryker, and supplies and set off to search for and explore the mysterious forested land that Herjolfsson claimed lay in a westerly direction. Some of the sagas even suggest that Erikson bought or borrowed Bjarni Herjolfsson's longboat to use on his epic voyage. When Leif Erikson set sail from Greenland, he earned his place in the history books as one of the greatest Norse explorers and the first European to set foot on the North American continent.

Even though Leif and his crew set sail in the summer, crossing the Atlantic Ocean so far north in an open longboat would have been a treacherous undertaking, and the crew would most likely have been plagued by high seas and ice flows. But these harsh conditions did not deter them. Leif and his crew were brave men and experienced seafarers, and they would not have turned back easily. It is uncertain how long the voyage took them, but the Saga of the Greenlanders claims that Leif and his crew made landfall at three different sites. The first place that they are believed to have landed was an icy and inhospitable region Erikson named "Helluland." This was a place of flat stone and ice. From there, they sailed on until they came to a heavily forested stretch of coastline that they called "Markland." They did not, however, choose to establish any sort of settlement there and sailed further south along the coast, quite possibly looking for a more suitable place to build a base camp. Approximately two days later, they came to a headland with an island just offshore. This appeared to be a more hospitable area than Helluland and Markland, and Leif decided to build his camp there and named the area Vinland.

The most common explanation for the name is apparently due to the abundance of wild grapes that the Vikings found growing in the area. But even this is open to interpretation and is still being debated by historians and scholars around the world. Some historians argue that the name cannot refer to grapes because if the settlement was far north then there would not have been wild grapes growing in abundance. Others claim that the grapes that are referred to in the name and apparently grew wild in Vinland may have been redcurrants or berries. In Scandinavian countries, there is a tradition of making wine by fermenting berries. However, if the Norsemen had traveled as far south as the St. Lawrence River and parts of New Brunswick, then they may indeed have found wild grapes. Another suggestion is that the name actually means "land of meadows" or "land of pastures."

There is also much debate as to exactly where Helluland, Markland, and Vinland were. Helluland means "Land of the Stones" or "Flat-Stone Land," and many historians now believe that this was most likely Baffin Island. The second place Leif and his crew made landfall they named Markland, which means "Woodland," and this is quite possibly Labrador. The exact location of the third landing place, Vinland, remains controversial, and it could have been as far north as Newfoundland or as far south as Cape Cod.

What is known is that the land that Leif and his crew found was far more hospitable than their native Greenland or Iceland. The ground was fertile, and there was abundant game to hunt. There was ample grazing, and the salmon and other fish in the rivers and lakes were large and plentiful. Most likely due to these favorable conditions, Leif and his crew decided to winter in Vinland and set about establishing a small settlement in a place they referred to as Leifsbudir (Leif's camp). They harvested timber not only to build their houses but also to take back to Greenland. The Norsemen were undoubtedly surprised by the mildness of the winter compared to what they were used to in Greenland. With the fertile soil and mild climate, it would have been possible to grow crops in the area if the Norsemen wanted to and being frost-free meant that it would not have been necessary to grow and store fodder for livestock in the winter. Legend has it that it was Thyrker who first discovered grapes on Vinland. In the spring, Leif and his crew returned to Greenland with a boatload of timber and grapes.

On his return voyage, Leif and his crew are said to have rescued the crew and salvaged the cargo of a trading vessel that had run aground on some rocks. This earned him the nickname "Leif the Lucky."

Leif only spent only one winter in Vinland before returning to Greenland with his precious cargo of timber and grapes. According to historical records, this is the only trip that Leif ever made to Vinland, and it is not known why he never returned to the rich and fertile North American continent.

What is known is that his father died around this time, either while Leif was still in North America or shortly after his return to Greenland. Whether this influenced his decision to not return to

Vinland we do not know. Once Erik and his followers had established a settlement on Greenland, they were followed by other migrants who helped expand and build the colony. However, one such group of immigrants, who arrived in 1002, brought with them an **epidemic** that ravaged the colony, killing many of its leading **citizens**, including the mighty man himself, Erik the Red. When Erik died in 1003, Leif became chief of the Eastern Settlement and lived Brattahlid for the rest of his life. When Leif died in 1025, his son, Thorkel Leifsson, became chief of the Eastern Settlement. After that, there is no further mention of any of Leif's descendants in history.

Leif and Thjodhild may have converted to Christianity, but when Erik died he was still a pagan who worshipped his Norse gods, and it is most likely that he received a pagan burial.[6]

[6]*It is impossible to know what Erik expected the afterlife to be like, as Norse paganism does not appear to have any consistent notion of the fate of the dead. However, most Vikings did believe in some kind of Land of the Dead and the most famous is undoubtedly Valhalla (the Hall of the Fallen), the hall of the god Odin. Little is known or recorded about the other Lands of the Dead but the goddess Freya is reputed to have welcomed some of the dead to her hall, Folkvang (the field of the people). Some, but not all, Vikings who died at sea were believed to reside in the underwater realm of the giant Aegir and his wife, Ran. The most common Land of the Dead or afterlife world in Norse paganism is Hel, an underground world presided over by a goddess named Hel. Vikings believed that here they would eat, drink, fight, sleep, and do most of the things that they had done while they were alive. Snorri Sturluson claimed that Viking warriors who died in battle went to Valhalla, and those who died under more peaceful circumstances were destined for Hel. This is an overly simplistic explanation of their beliefs, and one has to bear in mind*

that Snorri was writing as a Christian at a time when Norse paganism was no longer a living tradition. This simple explanation was most likely Snorri's interpretation of the Norse Lands of the Dead.

Chapter 7 – Subsequent Voyages to Vinland

Leif only made one voyage to Vinland but his discovery of North America assured his place in history as the first European to establish a settlement on the continent. Leif, himself, may not have returned to Vinland but his was not the one and only voyage that the Greenlanders made to North America. The settlement Leif and his crew had established on the continent remained, and more Greenlanders made the voyage across the ocean to settle, albeit temporarily, in this new land.

After Leif's historic voyage to Vinland, there were a number of other expeditions. These may have been merely to harvest timber and grapes or to try and establish a more permanent settlement on the continent. Erik the Red's courage and pioneering spirit appears to have influenced all his children, and they all followed in Leif's footsteps and attempted to make the voyage to Vinland at one time or another. Whether they were driven by a desire to emulate their father and establish a new settlement, to find a more hospitable and moderate climate to settle in, or merely to harvest timber, we will never know. Unfortunately, not all their voyages were successful.

The first of Leif's siblings to make the treacherous Atlantic crossing was his older brother, Thorvald. Thorvald used Leif's boat and followed his directions to make his way to Vinland. He spent approximately two years sailing along the coast, exploring the land, and harvesting timber and grapes. But of course, this was not an uninhabited territory, and the more time the Greenlanders spent on the North American continent the more they came into contact with the native population (called SKRÆLINGS by the Norsemen). They would most likely have traded with some of the Native American tribes they encountered, but more contact also led to conflict and skirmishes between the Greenlanders and local tribes. Thorvald Erikson was eventually killed in a skirmish with Native Americans and is the first European known to have died and been buried on North American soil. His crew buried him at a place they named Crossness. His crew did not remain in North America for long after Thorvald's death and returned to Greenland with timber and grapes.

After Thorvald's death, Leif's younger brother, Thorstein tried to make the crossing to Vinland in order to take Thorvald's body back to Greenland. Unfortunately, his voyage was hampered by fierce storms and high seas, and he was unable to reach the shores of Vinland and had to return to Greenland without his brother's body.

Thorfinn Karlsefni, an Icelandic trader, led another expedition to Vinland. Thorfinn, along with his wife, Gudrid, stayed in Vinland for three years and their son, Snorri, was born there. This made Snorri the first European child to be born on the North American continent. Unfortunately, relations between the Icelandic settlers and the local tribes were hostile, and eventually they abandoned their settlement and returned to their homeland.

Leif's sister, Freydis led the last recorded expedition to Vinland. According to the Greenland Saga, Freydis traveled to Vinland with two Icelandic traders and their crews, and she had her men kill the Icelanders before they returned to Greenland. No explanation is given for her actions. This is the last recorded expedition of Greenlanders to Vinland, but Icelandic records do mention an attempt by an Icelandic bishop, Eric Gnupsson, to visit Vinland from Greenland in 1121. Nothing more is written about Eric Gnupsson, and three years later a new bishop, named Arnald, is sent to Greenland. In 1347, reference is made in Icelandic records of a ship that arrived in Iceland with a boatload of timber from Markland. Apparently, the ship had been blown off course on its way home to Greenland. This record indicates that even though the Greenlanders abandoned their settlement on Vinland, they may have continued to harvest timber from the North American continent.

Chapter 8 – Master shipbuilders

Leif Erikson was clearly a remarkable man who was raised to be brave and adventurous. He was also obviously a master seafarer and navigator and these skills would have enabled him to cross the often-treacherous waters of the Atlantic Ocean and sail south along the coast of the North American continent until he found a suitable place to build a settlement. But it was not just his character and skills that made the voyage possible. Without the experience and incredible craftsmanship of Norwegian master shipbuilders, he would never have had the tools to cross the ocean and complete his historic voyage to Vinland. Norse shipbuilding was advanced for the age and the longboats these men designed and built enabled the Vikings to reach a height of naval power unheard of before in the region. Their longboats allowed them to conquer large areas of Europe, Britain, and the Middle East and ensured that their culture and influence spread across the known world. The typical longboat was wide and stable with a shallow draught that enabled it to travel upriver and land on beaches, but it was also a good seagoing vessel. The boats were light and fast, and there was nothing to rival them at the time.

The longboats had many uses. They were used to transport troops to battles and could also be tied together to form floating platforms for offshore battles. During the ninth century, the longboats played a pivotal role in the age of Viking expansion. With these remarkable boats, the Vikings were able to travel upriver and attack inland towns and cities such as Rouen in 841 and Hamburg in 845. The Vikings' enemies often referred to the longboats as "dragon ships" because of their dragon-shaped bow.

While the longboats are most famous for the role that they played in Viking raids and attacks, these boats were more than just warships. A longboat could be used in various ways and for a variety of missions. They could be used for fishing, seal and whale hunting, or to transport livestock, goods, and people over vast stretches of open water. The longboats were vital for trade and the exploration of new territories.

There were a number of different types of longboats, and mostly they were classified by the number of rowing positions on board. The Karvi was the smallest Viking longboat and had between 6 and 16 benches. This was a general-purpose vessel that was used for fishing and trade but could also be used in battle. The Snekkja had at least 20 rowing benches and could carry a crew of approximately 41 men. Snekkjas were useful in raids because they were light and could be beached or easily carried across portages. Skeids were larger warships and had more than 30 rowing benches and could carry around 70 to 80 men.

In 1926, Norwegian shipbuilders in Korgen built a replica of the type of longboat that Leif Erikson would have used on his journey to Vinland. The ship, named the Leif Erikson, was 42-foot-long and Captain Gerhard Folgero and his crew sailed it from Bergen, Norway, to North America. The ship stopped at the Shetland Islands, Faroe Islands, Iceland, and Greenland before crossing the Atlantic. On 20 July 1926, they docked at St. John's, Newfoundland. From there, they sailed south along the coast to Boston, Massachusetts, and finally reached Duluth in Minnesota on 23 June 1927. It was not an easy voyage, and the ship encountered heavy seas and even became ice-locked near Greenland, but this voyage proved that the Vikings were more than capable of sailing from Greenland to North America with their longboats.

Another useful tool in the Viking seafarers' arsenal was their exceptional navigational techniques. These skills allowed them to explore the ocean and discover new lands. They could travel far from land, establish trading posts in new territories and settlements in areas that had until then been inaccessible or hard to reach. While Viking navigational techniques are not well understood, it is clear that they were experts at judging speed, currents and wind direction and predicting tides. Norse traders regularly traveled between Norway, Iceland, and Greenland and, therefore, it was not unrealistic for them to attempt the crossing to North America.

But sailing over vast distances in open longboats took more than just remarkable shipbuilding ability, navigational skills, and seamanship, all of which the Norsemen clearly had in abundance; it also took courage. Traversing the deep sea was not for the faint-hearted. Ice on the rigging that could add crushing weight to the longboats was a constant danger in the northern seas. The men and women who made these voyages were remarkable people. They had the courage to sail across treacherous open waters and brave storms, fog, and ice to settle new lands. Women and children traveled alongside their husbands and fathers to colonize new lands and endured great hardships, pain, and deprivation without complaint. They too would have had to have a great deal of courage and resourcefulness in order to survive these voyages and carve out a new life for their families in harsh and unforgiving environments.

These were not people who were afraid to strike out on their own and make new lives in unknown lands, but in order to do so, they needed strong and competent leadership. It is clear that both Erik the Red and his son, Leif Erikson, possessed such qualities. Both were able to persuade people to join them on perilous voyages into the unknown, and this is an indication of the kind of men they must have been. They obviously inspired confidence in others, and people trusted them to make good decisions and act in the best interest of the group. Leading an expedition to unknown lands would have been a great responsibility, but it is clear that Leif Erikson was up to the task. He was able to navigate his way across the Atlantic to North America, establish a settlement, albeit temporary, and return to Greenland with a cargo of timber and grapes. He was also chosen as his father's successor in the Eastern Settlement and must have possessed good leadership skills.

Chapter 9 – The discovery of L'Anse aux Meadows

Most historians agree that Leif Erikson was the first European to set foot on North American soil and that the Norseman and his crew established some sort of settlement on this vast continent. The exact nature of this settlement, however, remains unclear and for various reasons, the Greenlanders never managed to establish a permanent and lasting settlement in this new territory. No records exist from this time, and the main source of historical information regarding the Norse voyages to Vinland is based on the Saga of the Greenlanders and the Saga of Erik the Red or on historical and archaeological finds in Greenland and North America. One of the difficulties that scholars face when interpreting the sagas is that both are based on oral histories and were only written approximately 250 years after the events they describe. These sagas talk about Leif Erikson's voyage to Vinland and describe the land he found, but they do not indicate exactly where Leif and his crew spent the winter on the North American continent. There are also no maps that survive from this time, and archaeologists and historians have to rely on secondhand descriptions to try and pinpoint the exact site where Leif settled. Due to this lack of accurate historical information, uncertainty still exists regarding the exact location of Vinland, and there are various theories about where Leif and his crew spent the winter on the North American continent.

In the early 1960s, a Viking site was discovered at L'Anse aux Meadows on the northernmost tip of the Northern Peninsula of Newfoundland in Canada. The excavation of L'Anse aux Meadows by Helge Ingstad and his wife, Anne Stine Ingstad, between 1961 and 1968 contributed greatly to the understanding of the Viking settlement on the North American continent. The Ingstads organized seven expeditions to L'Anse Aux Meadows where they identified eight or nine house sites similar to the ones found in Iceland and Greenland. The architecture, construction methods, and materials used for the houses are all similar to those used in early buildings in Iceland and Greenland and have been dated to the eleventh century. This would coincide with the sagas and the time period in which Leif would have made his voyage to Vinland. Ingstad and his wife also found an assortment of Viking artifacts, including a ring-headed pin, soapstone spindle whorl, and a smithy with a large stone anvil. The iron that was excavated at the L'Anse aux Meadows site was produced from bog-iron using the same methods used in Norway and Iceland during the Viking Age.

The Viking settlement discovered at L'Anse aux Meadows is certainly similar in many ways to sites in Iceland and Greenland, but it also differs in a few significant aspects. In Iceland and Greenland, Viking settlements were usually in sheltered areas, but L'Anse aux Meadows is more exposed to the elements. There also don't appear to be any barn-like structures, enclosures, or shelters for livestock, but there were large storage rooms where they could have stored timber and grapes before taking them back to Greenland.

When Ingstad excavated the Viking settlement at L'Anse aux Meadows, he was convinced that he had found Vinland and his discovery provided irrefutable proof that the Icelandic sagas were true and Leif's voyage did take place. But while historians and scholars no longer dispute the fact that Leif Erikson and various other Vikings led expeditions to the North American continent, not all are convinced that L'Anse aux Meadows is indeed Vinland. One school of thought theorizes that L'Anse Aux Meadows could have been a repair station used by Leif and those who came after him as a place to repair their longboats after the harsh Atlantic crossing. They would have stopped there for a while before they sailed further south to Vinland. Until more evidence of Viking settlements is found in North America, the controversy regarding the exact site of Leif's Vinland will remain.

Even the meaning of the name Vinland has caused debate among historians. As mentioned earlier, the name can be interpreted in various ways. If one takes Vinland to mean "Wine Land" then some scholars believe that L'Anse aux Meadows is too far north to be the area Leif referred to as Vinland. While Vinland may not have been the land of milk and honey, it was described as a land of abundant grapes, fertile soil, and a moderate climate, and L'Anse aux Meadows appears to fall short of this description. There are certainly no grapevines growing in wild abundance that far north on the continent. But if, as Ingstad and his supporters suggested, one translates the "*vin*" in Vinland as meadow or pasture then he may well have discovered Leif's Vinland.

But if L'Anse aux Meadows is not the site of Leif's Vinland, then where could it have been located? If one interprets Vinland as meaning "Wine Land" then the Gulf of Saint Lawrence could well be the site of Leif's settlement. The Gulf of Saint Lawrence is approximately 700 nautical miles south of L'Anse aux Meadows and includes Prince Edward Island, Nova Scotia, and New Brunswick. This is an area that Leif may well have called Vinland due to the abundance of wild grapes that grow in the area. Unfortunately, no evidence of a semi-permanent Viking settlement has been found in the Gulf of Saint Lawrence.

The site at L'Anse Aux Meadows was declared a National Historic Site of Canada in the 1970s, and further excavations revealed over 2,000 Norse and Native American artifacts from various centuries. All the artifacts gathered at L'Anse Aux Meadows and the knowledge that the site has provided might not conclusively prove that this is the site of Vinland, but it does gives historians insight into Viking life on the North American continent.

Chapter 10 – The End of the Vinland Settlement

The land that Leif Erikson discovered appeared to offer the Greenlanders everything they were looking for. The climate was moderate, there was plenty of grazing for livestock, there were large forested areas, and the land was fertile. It was a far more hospitable environment than Greenland, and yet the Vikings failed to establish a permanent settlement on the continent. From the time of Leif's incredible voyage to the abandonment of the settlement was less than ten years.

So why did the Norsemen abandon their attempts to settle in North America? One reason could be that the resources Vinland had to offer were not attractive enough to make the dangerous voyage worthwhile. The distance between Greenland and Vinland was almost 3,500 kilometers, and that was a long voyage to make just for timber and grapes when the same commodities could be obtained in Norway which was a far shorter and less treacherous journey to make regularly.

Vinland had very little to offer the Greenlanders when compared with Europe. From Europe, the Greenlanders could get spices, salt, metals, textiles, glass, and other luxuries that were not available in Vinland, and they could build political and religious connections. Regular voyages between Greenland, Iceland, and Norway were a necessity for the Greenlanders to ensure their survival as a colony, but Vinland offered no such advantages. A voyage to Vinland was a dangerous and expensive expedition without any guaranteed rewards. One example of a failed expedition was Thorstein's voyage that never even reached the shores of Vinland. He spent almost the entire summer sailing in the Atlantic Ocean and returned empty handed. A small and newly established colony like Greenland could not afford to have 30 to 40 of their able-bodied men sailing aimlessly in the Atlantic and returning with nothing. These men's skills could be better used at home. Simply put, the resources available in Vinland were not worth the effort it took to get them to Greenland, and a voyage to Vinland offered almost no advantages over a voyage to Europe.

Viking longboats, as advanced and sophisticated as they were for their time, were not ideal oceangoing vessels. They were not originally designed for long voyages across the open sea, and though it was possible to make the voyage, as Leif proved, it was not necessarily practical. Because of the limitations of the longboats, it was almost impossible and certainly impractical to bring settlers and supplies from Norway or Iceland all the way to Vinland. This meant that potential settlers had to come from Greenland, and as it was a relatively new and barely settled territory with very few resources, this was not really a viable option. Yes, it was possible for Leif Erikson and a few other Greenlanders to make exploratory voyages across the Atlantic, but there was no way that the Greenlanders could support and sustain a satellite settlement so far from their shores.

The expense of maintaining a permanent colony on the North American continent would have been a crippling drain on Greenland's already limited resources and manpower. Having too many of their able-bodied men across the Atlantic harvesting timber and grapes would have put enormous pressure on those who remained in Greenland.

Unlike Greenland, North America was already inhabited by a large indigenous population when the Vikings arrived there. The Vikings did trade with the local tribes but almost from the start there were hostilities between the Norsemen and the Native Americans. During his three years in Vinland, Knarlsefni realized that the land might have much to offer, but the Norsemen would never be safe there, and they would live under constant threat from the local tribes. The number of settlers was relatively small and therefore vulnerable to attack. Unlike Christopher Columbus and his men, the Norsemen did not have weapons that were superior to those of the local inhabitants. The distance between Vinland and the Viking strongholds of Greenland and Iceland was just too vast for them to defend their location and settlers from attacks by Native American tribes.

Besides the fact that there were very few pull factors encouraging the Vikings to settle permanently in Vinland, there were also very few push factors. When Erik the Red had left Iceland to settle in Greenland, he found many followers because Iceland was becoming overpopulated, and many of those who left were looking for a better life and more freedom. At the time of Leif's voyage to Vinland, there was no population pressure pushing the Greenlanders to seek greener pastures elsewhere. At the time, Greenland was a relatively new settlement and not yet well established. There was an abundance of game to hunt, the fjords were well stocked with fish, and there was plenty of lush grazing for livestock in the summer.

There was also much work to be done, clearing the land and building homesteads.

The failure of the settlement in Vinland was at its heart a numbers problem. The distance between Greenland and Vinland was too great. The settlers were vastly outnumbered by the indigenous people, and conflict was growing between the two groups. Greenland also did not have the numbers in their own settlement to support another satellite settlement. It would have taken too great a percentage of Greenland's prime working population to support a permanent settlement so far away and harvest the natural resources available in Vinland.

This basically meant that once Leif and the others had been to Vinland, explored the area, and wintered there a few times, they left again and never returned. But this does not mean that Leif's incredible voyage was in vain. Leif left a lasting legacy on world history. Greenland would have been enriched by Leif's voyage and the timber and grapes that he brought back from North America. His discovery, although forgotten for many years, would have contributed to the history of the territories he explored and would have influenced explorers and seafarers who came after him. Norsemen continued to make sporadic voyages at least as far as Markland for timber. Knowledge of Leif's voyage to Vinland also spread to Europe, and writers such as Adam of Bremen mentioned the remote land in their writings. For all we know, Christopher Columbus might even have heard stories about Leif and his extraordinary voyage.

Chapter 11 – The Decline of Greenland Settlement

After Leif left the North American continent and returned home with his cargo of timber and grapes, he lived in Greenland for the rest of his life. He would no doubt have traveled to Iceland and perhaps even Norway, but he made no other historic voyages. Or if he did, they are not recorded anywhere. There is in fact very little information about Leif after his return from Vinland.

We know that Leif became chief of the Eastern Settlement when Erik the Red died, and it is assumed that he remained chief until his death and then the mantle passed to his son, Thorkel Leifsson. After that, the family of the famous explorer and his mighty Viking father fades into obscurity. This is mostly due to the lack of any historical records surviving from the Greenland settlement. Perhaps if the Greenland settlement had survived to modern day the incredible feats of Erik and Leif would have been recorded and celebrated by the local population, and their descendants would have been traceable on the island. But that is sadly not the case.

Unfortunately, in the same way that the Norse settlement at Vinland didn't survive, neither did the Greenland settlement. The land that Erik the Red, Leif, and the other colonizers had struggled so hard to tame eventually became too marginal to support their descendants. Approximately 500 years after Erik and his followers landed on the shores of Greenland, both the Eastern and Western settlements had been abandoned. Incidentally, this was around the same time that Christopher Columbus was making his historic voyage to the Americas. Theories and speculation abound as to why Greenland was abandoned, but the exact reasons for the decline are unknown. What is known is that sometime between the fifteenth and seventeenth centuries, the entire Norse population of Greenland vanished.

It is unlikely that one single catastrophic event ended the life of the settlement and more likely that a combination of events led the Greenlanders to abandon their homes and move to more hospitable environments. One contributing factor to the decline of the settlements could be the Little Ice Age. This was a period during which Europe and North America were subjected to exceptionally cold winters; the resulting crop failures and famine could have meant the end of the colony. Life would have become too hard for colonists as farming on Greenland became marginal at best. The Greenlanders would still have been able to survive by eating a seafood-based diet and may have shifted their focus from farming to hunting seals, fur trading, and fishing. But the Greenlanders were farmers at heart, and as the grazing became sparse, cattle were replaced by goats and sheep. Life on Greenland was always hard, but maybe the change in climate meant that the population felt it was just not worth the daily struggles anymore and moved on to more hospitable climates.

Their economy may also have declined due to changing fashions and less demand for the resources that the Greenlanders had access to, including their two biggest exports: walrus tusks and seal skins. There was also no longer regular ship traffic between Iceland and Norway and this would have severely impacted the market that the Greenlanders had for their resources. The voyage from Norway to Greenland was long and treacherous, and few traders were willing to travel that far from the mainland of Europe, and this meant that supplies to the colony became more sporadic and more expensive.

Isolation probably also played a role in the decline of the colony. Increasing isolation meant that the Greenlanders were losing touch with their national identity. Without regular trade and travel to Norway, the Greenlanders were becoming more and more isolated from their homeland and culture, and this would have impacted the mental health of the colony.

Whether it was disease, famine, climate change, attacks by raiders, loss of trade and identity, or the harsh environment and daily struggle for survival, or a combination of these factors that led to the decline of the Greenland settlement, we will probably never know. As is the case with most emigration, it was probably the younger generation that abandoned the colony first. Strong, able-bodied Greenlanders of child-bearing age would have moved to areas where more opportunities existed for them. The abandonment of Greenland seems to have been an orderly process rather than a mad scramble. Excavations on the island have turned up very few valuables, and archeologists believe this means that the settlers did not leave in a hurry and took everything of value with them. Besides not knowing why the Greenlanders left, there are also no records of where they went, and they were probably assimilated back into Norwegian society.

Greenland may never reveal all its secrets, and the reasons for the decline of this colony may be lost, but the relatively brief time that the Norsemen spent on this harsh and inhospitable land played a vital role in the Viking Age of exploration and expansion. Without the colony on Greenland, Leif Erikson may never have made his epic voyage to North America, and Christopher Columbus might well have been the first European to set foot on this vast continent. The settlement at Greenland not only made the physical voyage to North America possible, but it also made Leif Erikson the man he was. He and his siblings were clearly a product of their upbringing and environment. They inherited their father's adventurous and pioneering spirit, and the harsh land that they grew up in gave them the skills they needed to venture into the unknown and attempt to establish a colony far from home. Sailing regularly between Greenland and Iceland and traversing the Norwegian Sea would undoubtedly have developed Leif's seafaring ability and navigational skills to a point where he had the confidence to sail across the treacherous waters of the Atlantic to look for the mysterious land that Bjarni claimed lay to the west.

Chapter 12 – Leif Erikson's place in History

Leif Erikson is widely recognized and commemorated as the first European to set foot on the vast North American continent. He landed there almost half a century before Christopher Columbus, but for many years his story was all but forgotten. Children in American schools were taught that Christopher Columbus was the first European to discover America, and they still celebrate Christopher Columbus Day. But during the nineteenth and twentieth centuries, Leif Erikson reclaimed his rightful place in history and today, statues commemorating this incredible Norse explorer and his contribution to the world can be found in various countries.

In 1887, a statue sculpted by Anne Whitney was erected in Boston, Massachusetts, and two weeks later, a replica of the statue was erected in Milwaukee, Wisconsin. In 1962, a sixteen-foot statue by August Werner was erected in Seattle, Washington, and today there are more than half a dozen statues of Leif Erikson throughout America. There are also replicas of the Seattle statue in Trondheim in Norway, Brattahlid in Greenland, and L'Anse aux Meadows in Newfoundland.

Leif Erikson is now also celebrated in America and October 9 is Leif Erikson Day. In 1925, President Calvin Coolidge recognized Leif

Erikson as the first European to visit North America, and in 1929, Wisconsin became the first state to officially make Leif Erikson Day a state holiday. In 1931, Minnesota followed suit, and by 1956, Leif Erikson Day was observed in seven states and one Canadian province. In 1963, John Blatnik, the US Representative from Duluth, introduced a bill to observe Leif Erikson Day nationwide, and in 1964 it was adopted by Congress. From then on, every president has issued proclamations about the holiday and has often used the opportunity to publicly praise the Norse spirit of exploration and discovery and to highlight the contributions the Scandinavians have made to American culture.

The date chosen for Leif Erikson Day has no particular connection to any events in his life but is the day in 1825 that the ship, Restauration, carrying immigrants from Norway, arrived in New York. This was the start of a wave of organized immigration from Norway to the United States. Over the next hundred years, nearly one-third of the Norwegian population immigrated to the United States, and there are more than 4.5 million American with Norwegian ancestry living in America today. Leif Erikson Day is, therefore, not just an opportunity for Americans to celebrate the remarkable achievements of Leif Erikson but also an opportunity to celebrate their Nordic heritage.

One of the key differences between Leif Erikson and Christopher Columbus was that Erikson never intended to colonize the North American continent. He was not an empire builder or acting on behalf of a European monarch. What Leif tried to establish in Vinland was most likely never intended to be more than a satellite settlement of the Greenland colony. Leif probably saw Vinland as nothing more than a source of resources, such as timber and grapes that were scarce or non-existent on Greenland.

Timeline of Leif Erikson's life

The exact dates of Leif Erikson's birth, death, and voyage to the North American continent are unknown, but a timeline can be drawn up that approximates the most important dates in his life.

Around 970 – Leif Erikson is born in Iceland.

Early 980s – Erik the Red is banished from Iceland for three years for the murder of Eyiolf the Foul. During his exile, Erik explores Greenland. On his return to Iceland, he convinces his family and a group of Vikings to travel to Greenland and establish a permanent settlement.

986 – Icelandic explorer Bjarni Herjolfsson is blown off course on his way to Greenland and sights land to the west.

999 – Leif Erikson sails to Norway. On the way he stops at the Hebrides where he fathers a son, Thorgils, with Thorgunna, the daughter of a local chief. In Norway, Erikson spends the winter in the household of King Olaf Tryggvason and is converted to Christianity.

1000 – Erikson returns to Greenland. Aware of Bjarni Hefjolfsson's tales of a mysterious land to the west of Greenland, Erikson and a small group of Vikings set sail to discover this land. Leif and his crew establish a small settlement in an area that he calls Vinland and which historians believe could be modern-day Newfoundland.

1003 – Leif leaves the settlement at Vinland and returns to his father's estate at Brattahild in Greenland. Shortly before or after his return, his father dies and Leif becomes chief. He never returns to Vinland.

Between 1004 and 1010 – The Greenlanders make more voyages to Vinland. Leif's brother, Thorvald, is killed by the local inhabitants in North America and is the first European to die on the continent. Snorri Thorfinnsson, the son of Thorfinn Karlsefni and Gudrid, is born in Vinland. He is the first European child born on the continent.

1010 – The settlement at Vinland declines and fails mainly due to attacks from indigenous tribes and unsustainability.

1025 – Leif Erikson dies and his son, Thorkel Leifsson, becomes chief of the Eastern Settlement.

Conclusion

Unfortunately, due to the lack of accurate historical records and limited archaeological finds, the impact that Leif Erikson had on world history can never be truly quantified. But Leif Erikson was, without a doubt, an extraordinary man, and he deserves his place in history as a renowned explorer and the first European to set foot on the North American continent. This mighty Norseman had the courage and vision to leave the safe shores of his homeland and discover America almost half a century before Christopher Columbus. This took remarkable seafaring ability and navigational skills. He crossed the treacherous Atlantic Ocean in a Viking longboat with a crew of 35 men and redefined the boundaries of the world he lived in. His successful voyage to Vinland encouraged others to follow in his footsteps and visit the North American continent, and contact with the local tribes would have undoubtedly had a lasting effect on both cultures. Leif also had the power and influence to spread Christianity to Greenland and convert many of the Greenlanders to the Christian faith. This is a man worth celebrating, not just for his deeds, but for his courage and determination.

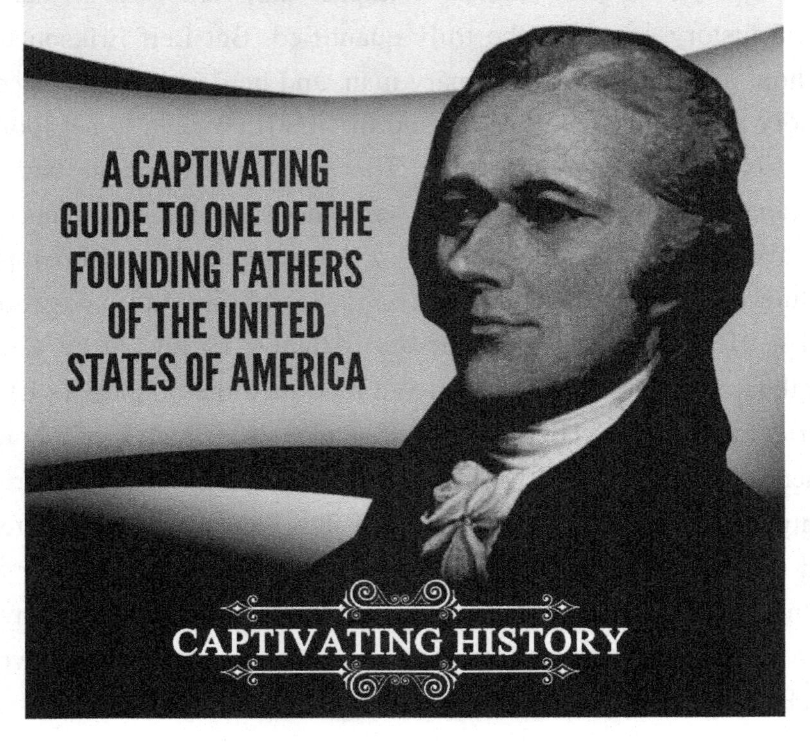

Check out this book!

Check out this book!

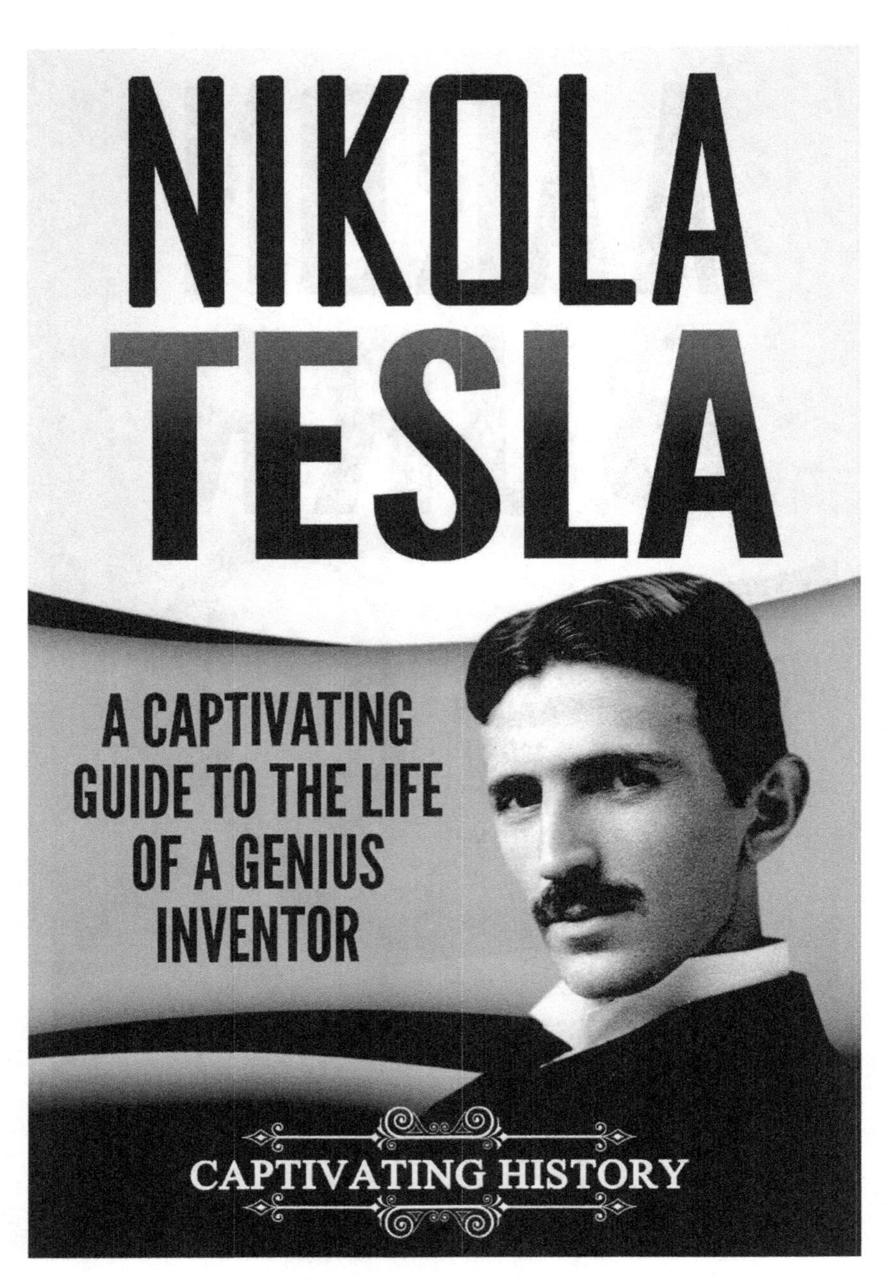

Check out this book!

THOMAS EDISON

A CAPTIVATING GUIDE TO THE LIFE OF A GENIUS INVENTOR

CAPTIVATING HISTORY

Check out this book!

Free Bonus from Captivating History (Available for a Limited time)

Hi History Lovers!

Now you have a chance to join our exclusive history list so you can get your first history ebook for free as well as discounts and a potential to get more history books for free! Simply visit the link below to join.

Captivatinghistory.com/ebook

Also, make sure to follow us on:

Twitter: @Captivhistory

Facebook: Captivating History:@captivatinghistory

Bibliography

www.viking.no

www.leiferikson.org

www.wikipedia

en.natmus.dk (National Museum of Denmark website)

History.howstuffworks.com

www.history.com

www.bbc.com/history/historic_figures/erikson.leif

www.britannica.com/biography/Leif.Eriksson-the-lucky

Mentalfloss.com

www.biography.com/people/leif-eriksson

www.ancient-origins.net

www.timeanddate.com/holidays

www.wincalander.com/Leif-Erikson-Day

www.canadiannmysteries.ca

Danver, Steven (2010); Popular Controversies in World History: Investigating History's Intriguing Questions

Ingstad, Helge; Ingstad, Anne Stine (2001) The Viking Discovery of America: The Excavation of a Norse Settlement in L'Anse aux Meadows, Newfoundland

Wallace, Birgitta; The Norse in Newfoundland: L'Anse aux Meadows and Vinland

Wallace, Birgitta; L'Anse aux Meadows and Vinland: An Abandoned Experiment

L'Anse aux Meadows National Historic site of Canada

A.M. Reeves et al; The Norse Discovery of America

The Icelandic Saga Database: The Saga of Erik the Red (sagadb.org/eiriks_saga_rauda.en.pdf)

Groenlendiga Saga (notendur.hi.is/haukurth/utgafa/Greenlanders.html)

D.J. Williamson et al; The Rules of Revenge; Viking Justice (All About History, Issue 059)

Davis, Graeme; The Cult of Thor (All About History, Issue 057)

Made in the USA
Las Vegas, NV
26 March 2025

20159062R00046